MOUNTAINS OF
SCOTLAND

Colin   land

MOUNTAINS OF SCOTLAND

Almost half a billion years ago something happened which changed forever the face of Scotland. During the blink of an eye – in geological terms something like 20 million years – southern and northern continental land masses joined as the Iapetus Ocean between them closed. As a result, Scotland merged with what is now the rest of the British Isles. As the two continents collided, mountains formed, to be sculpted later by the elements of ice and snow.

Today you can stand on rocks which reach back to the beginnings of time. In the north-west Highlands you can touch rocks which lay thousands of feet deep in the earth three billion years ago. Then, they were part of the earth's crust. But as the crust broke up and drifted, and the elements of wind, rain and ice bore down on the exposed rocks, the mantle of the land took on its present-day appearance.

The highest mountain in Scotland, indeed Britain, is Ben Nevis 4409 ft, (1344 m). This is one of Scotland's 'Munros' – the 284 mountains higher than 3000 ft (914 m). Smaller hills above 2500 ft (762 m) are known as the 'Corbetts', of which there are 221. Braeriach, Cairn Toul and Ben Lawers rank among the highest ten mountains in Scotland; but many of our most striking hills are not even Corbetts, notably Suilven and Stac Pollaidh. Height alone tells us little. Rather, it is the colour, shape and wildness, which forms lasting impressions.

Ben Venue from Ben A'an, The Trossachs

Overlooking the Trossachs and Loch Katrine, Ben Venue lies some 6 miles
(10 km) east of Ben Lomond and lies at the heart of the romantic Highlands made famous
by Sir Walter Scott. Beneath its slopes, there are forests of birch, oak, ash and alder.

◄ Ben Lomond – The Most Southerly Munro

'THE COBBLER' – BEN ARTHUR, ARGYLL

One of Scotland's most popular and distinctive hills. The centre peak is the summit;
the mountain has some 80 rock climbs founded on the mica-schist rocks.

BEN LUI (BEINN LAOIGH) FROM BEN OSS, ARGYLL
– The Aptly Named Black Rock Hill ▶

BEN LAWERS, BREADALBANE – A Summit of Riches Seen Here from Kenmore

The highest mountain in the southern Highlands, with eight tops above 2950 ft (900 m). The upper rocks of the mountain escaped the ravages of the last Ice Age, and it is believed that this is why arctic-alpine plants grow so successfully here.

◄ BEN CHALLUM, BREADALBANE – At the Head of Glen Lyon

◀ GLENCOE –
The Spiritual Home of Mountaineering in Scotland

Bold, rugged, revered and poignant. Glencoe is dominated by the high Bidean massif in the south and the spectacular wall of the Aonach Eagach ridge to the north. In the darkness of 13 February 1692, 38 MacDonalds were murdered here, leaving a lasting legacy to the savagery of the Jacobite period. This region deserves a different reputation – as one of the most exciting, raw mountainscapes in Scotland.

BUACHAILLE ETIVE MÓR, GLENCOE

Towering over Glen Coe, and Glen Etive to the south, this is the epitome of the grand entrance to Scotland's mountains. Here, the experience is sharpened by the prelude of the great bog of Rannoch with its remnant pine stumps betraying a rather different, wooded landscape of the past.

BEN NEVIS FROM AN GARBANACH, LOCHABER

Ben Nevis stands proud over Scotland with its great crags, boulder fields and summit plateau. This is a fragile place where the soils yield readily to erosion from people and the weather. Conditions are arctic-like, with snow lasting until July and, on the tops, some snowbeds last almost year round.

BEN NEVIS
FROM CARN DEARG
MEADHONACH,
LOCHABER ▶

Rock climbs here were 'discovered' in 1892 by the Hopkinson family of northern England. You can picture the three brothers and a son toiling up the hill to meet what were later named Tower Ridge and the North-East Buttress.

This is a fascinating area for geologists and botanists, and there may still be plants here which when found will be new to Scotland. A Weather Station was maintained at the summit from 1883 for around 20 years.

◀ SCHIEHALLION,
LOCH RANNOCH, PERTHSHIRE
– The Fairy Hill of The Caledonians

This pyramid-like hill is very distinctive.
Climb to the summit and discover the
fine views of Ben Alder and the Forest
of Atholl.

LOCHNAGAR, ABERDEENSHIRE ▶

Named after Lochan na Gaire, this
mountain is famed for its associations
with the Royal Family. Bought by
Queen Victoria and Prince Albert in the
nineteenth century, it inspired Lord Byron
to write: *England! Thy beauties are tame
and domestic to one who has roved o'er the
mountains afar: Oh for the crags that are
wild and majestic! The steep frowning
glories of dark Lochnagar.*

◀ Lurcher's Gully and The Northern Corries of The Cairngorms

The Northern Corries are a supreme example of the nature of ice erosion. This unique landscape has features that are world renowned for displaying the natural forces of ice and other aspects of the climate, which shaped the corries, terraces and other splendid features of the mountains.

An Garbh Choire, Cairngorms – Between Braeriach and Cairn Toul ▶

An ethereal dawn testifying to the aftermath of the Ice Age which ended over 10,000 years ago. Here, the ice carved out the troughs and laid bare rock faces.

Loch Hourn and Knoydart, West Lochaber

The mountains of Knoydart brood over Loch Nevis and Loch Hourn
– the nearest imitation in Scotland to a Norwegian fjord.

The Mountains of Kintail, Glen Shiel ▶

This massive mountainous peninsula is one of the most impressive north of the
Great Glen. You can spend a week in these hills in complete isolation. The south face
of Sgurr na Ciste Duibhe – one of the 'Five Sisters' – is the steepest in Britain.

◀ THE BLACK CUILLIN, SKYE

Here are the remnants of a once-great
volcano. The rock is gabbro – the
solidified remains of molten rock from
the bubbling magma chamber which
erupted some 65 million years ago.
This is one of Scotland's youngest
mountain complexes, offering spectacular
rock and winter climbs; the main ridge
is almost 1000 metres high. The Gaelic
poet Sorley MacLean referred to these
mountains as *mur eagarra gorm* –
'the exact and serrated blue rampart'.

BLA BHEINN, SKYE
– The Blue or Warm Hill ▶

Known to some as Blaven, and the
attractive cousin of the Cuillin. There
are magical views from here, towards
the Cuillen Ridge to the west.

◀ Liathach, Torridon
  – Peak and The Gray Corrie

Viewed from the north-east this great rampart proclaims itself a majestic mountain with its great tiers of Torridonian sandstone and quartzite ridges and pinnacles.

Beinn Eighe, Torridon ▶

Britain's first National Nature Reserve. This mountain is famed for its old Scots pine woods and plant communities characteristic of Northern Atlantic regions. It is one of our most oceanic mountains, with over 220 wet days each year. The Cambrian quartzite ridges and Torridonian sandstone exposures are spectacular. Coire Mhic Fhearchair at the west end of the mountain is one of Britain's most magnificent corries.

◄ SLIOCH AND
  LOCH GARBHAIG,
  WESTER ROSS

Derived from the Gaelic *sleagh* meaning 'spear', Slioch is a grand rock fortress with an air of impregnability. This is the hallmark landscape of the north-west Highlands and forms part of a land which transcends billions of years. Some 800 million years ago Slioch would have been much higher and more extensive. But the passage of ice and erosion has left us with a much smaller hill.

AN TEALLACH, WESTER ROSS
  – The Forge, Peak of the Greenish-Grey Hollow

Viewed from the air, east of An Teallach, Loch Toll an Lochain is near frozen. This vista hints at the exhilarating experience to be had here. On the mountain you have spectacular views and raw, physical experiences as you scramble over rock and ridge. Adventure at its best.

STAC POLLAIDH, INVERPOLLY
– Peak of the Peat Moss

Part of Inverpolly National Nature Reserve,
this mountain is situated north of Loch Lurgainn.
It is accessible, rendering it highly vulnerable to erosion.
Its delightful shape stands proud over heaths and bogs,
appealing to the eye.

SUILVEN, SUTHERLAND
– The Grey Castle ▶

When first encountering
Suilven, the invincibility
of this mountain is evident.
Norsemen referred to it as
the 'pillar' as they looked
up from Lochinver. It is
an utterly magnificent
mountain.

You sneak a glimpse
of Suilven from much of
western Sutherland. It is a
challenging hike, with only
the southern and northern
flanks accessible on a 12 mile
(20 km) round trek from
Lochinver or from the east.
Memories of this mountain
last a lifetime.

STRATH DIONARD AND FOINAVEN, SUTHERLAND

Driving south from Durness you come across this fantastic strath, seeing
Beinn Spionnaidh (2533 ft, 772 m) first, and then the ridges of Foinaven (2989 ft, 911 m).

◄ BEN MORE ASSYNT AND CONIVAL, SUTHERLAND

A great massif best climbed from Inchnadamph, where the limestone
cliffs have caves and potholes adding to the adventure!

BEN HOPE, SUTHERLAND – Hill of the Speckled Cliff

Britain's most northerly Munro, Ben Hope offers a wonderful perspective
on the north of Scotland. Look out for the heavy clouds which cling to the summit!
The high plateaux resemble some of the desolate arctic plains of North Scandinavia.

BEN LOYAL AND THE KYLE OF TONGUE, SUTHERLAND ▶

BEN CRUACHAN, LORN – The Stacky Hill, Peak of The Stag

Published in Great Britain in 2002 by
Colin Baxter Photography Limited,
Grantown–on–Spey, Moray PH26 3NA, Scotland.
www.colinbaxter.co.uk

Text by Des Thompson, Copyright © Colin Baxter Photography Ltd 2002
All photography Copyright © Colin Baxter 2002

A CIP catalogue record for this book is available from the British Library.
ISBN 1-84107-123-4   *Colin Baxter Gift Book Series*   Printed in Hong Kong

Front cover photograph: Above Ben More, near Crianlarich, Perthshire.
Page one photograph: Ben More, Isle of Mull.
Page two photograph: Ben Nevis – Britain's highest mountain.
Back cover photograph: The Black Cuillin, Isle of Skye.